math standards workout

PROBLEM
SOLVING
50 MATH SUPER PUZZLES

By Thomas Canavan

This edition first published in 2012 by The Rosen Publishing Group, Inc.
29 East 21st Street, New York, NY 10010

Copyright © 2012 Arcturus Publishing Limited

Author: Thomas Canavan
Editor: Joe Harris
Design: Jane Hawkins
Cover design: Jane Hawkins

Library of Congress Cataloging-in-Publication Data

Canavan, Thomas, 1956-
Problem solving : 50 math super puzzles / Thomas Canavan.
p. cm. — (Math standards workout)
Includes bibliographical references and index.
ISBN 978-1-4488-6675-5 (library binding) — ISBN 978-1-4488-6682-3 (pbk.) — ISBN 978-1-4488-6688-5 (6-pack)
1. Problem solving—Juvenile literature. 2. Mathematical recreations—Juvenile literature. I. Title.
QA63.C35 2012
513—dc23
2011028574

Printed in China
SL002076US

CPSIA Compliance Information: Batch #W12YA. For further information, contact Rosen Publishing, New York, New York, at 1-800-237-9932.

Contents

Introduction

Why do you need this book?

Do you feel that your math muscles could benefit from a good workout? Each of the 50 puzzles in this book is a challenge to your problem-solving skills. After a good workout, you probably feel that your weaker skills got a little stronger—and your real talents got that much sharper. The same is true here: this book will help you to build on your problem-solving strengths and improve in those areas where you are less confident.

How will this book help you at school?

Problem Solving complements the National Council of Teachers of Mathematics (NCTM) framework of Math Standards, providing an engaging enhancement of the curriculum in the following areas:

> *Problem Solving Standard for K—12*
> *Geometry: Use Visualization, Spatial Reasoning, and Geometric Modeling to Solve Problems*

Why have we chosen these puzzles?

This *Math Standards Workout* title features a range of interesting and absorbing puzzle types, challenging students to master the following skills to arrive at solutions:

- Build new mathematical knowledge through problem solving: e.g. Number Crunch, Pyramid Plus

- Solve problems that arise in mathematics and in other contexts: e.g. Making Arrangements, Mini Sudoku

- Apply and adapt a variety of appropriate strategies to solve problems: e.g. One to Nine, Hexagony

- Monitor and reflect on the process of mathematical problem solving: e.g. Number Path

NOTE TO READERS

If you have borrowed this book from a school or classroom library, please respect other students and DO NOT write your answers in the book. Always write your answers on a separate sheet of paper.

Number Crunch

Starting at the top left with the number provided, work down from one box to another, applying the mathematical instructions to your running total. Write your answers on a separate sheet of paper.

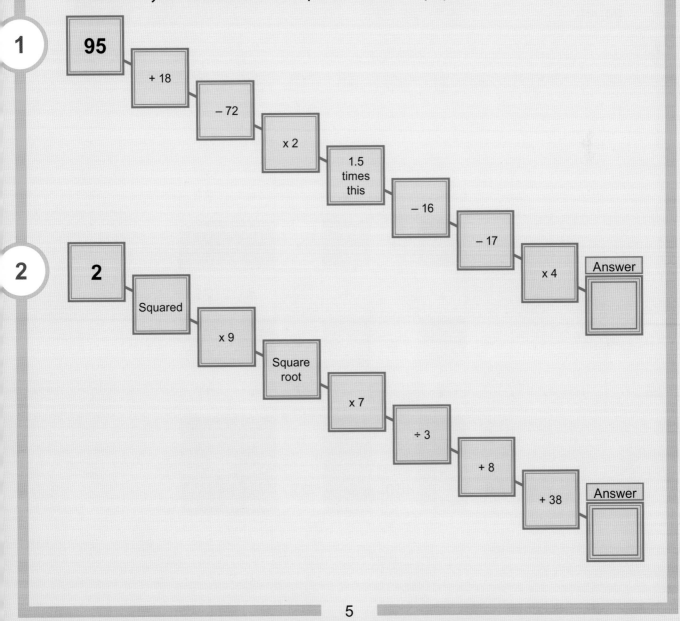

1

95

+ 18

− 72

x 2

1.5 times this

− 16

− 17

x 4

Answer

2

2

Squared

x 9

Square root

x 7

÷ 3

+ 8

+ 38

Answer

One to Nine

Using the numbers one to nine, complete these six equations (three reading across and three reading downward). Every number is used once only, and one is already in place. Write your answers on a separate sheet of paper.

3

1 2 3 4 5 6 7 8 9

	−		x		=	35
+		+		x		
	x		+		=	21
−		x		+		
	x		−	7	=	17
=		=		=		
6		60		22		

Making Arrangements

Arrange one each of the four numbers below, as well as one each of the symbols x (times), – (minus), and + (plus) in every row and column. You should arrive at the answer at the end of the row or column, making the calculations in the order in which they appear. Some are already in place. Write your answers on a separate sheet of paper.

4

Total Concentration

The blank squares below should contain whole numbers between 1 and 30 inclusive, any of which may occur more than once, or not at all. The numbers in every horizontal row add up to the totals on the right, as do the two long diagonal lines extending from corner to corner; those in every vertical column add up to the totals along the bottom. Write your answers on a separate sheet of paper.

							138
18		3		2	6	22	93
2	3	17	11			11	65
12		19	10	13	5	29	111
	9	10	30	19			119
7	26		13		8		76
	14	1	5	16		8	63
27	20			9	21	28	128
82	113	83	100	66	95	116	114

Mini Sudoku

Every row, column, and each of the four smaller boxes of four squares should contain a different number from 1 to 4 inclusive. Some numbers are already in place. Can you complete the grid? Write your answers on a separate sheet of paper.

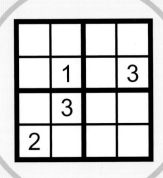

Every row, column, and each of the six smaller boxes of six squares should contain a different number from 1 to 6 inclusive. Some numbers are already in place. Can you complete the grid? Write your answers on a separate sheet of paper.

Pyramid Plus

The number in each circle is the sum of the two numbers below it. Just work out the missing numbers in every circle! Write your answers on a separate sheet of paper.

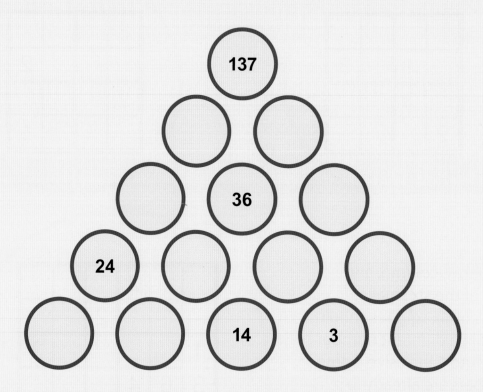

Tile Twister

Place the eight tiles into the puzzle grid so that all adjacent numbers on each tile match up. Tiles may be rotated through 360 degrees, but none may be flipped over. Write your answers on a separate sheet of paper.

10

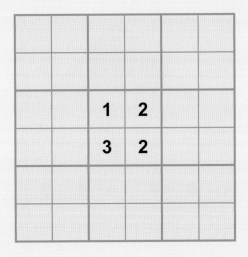

2	1
1	3

3	1
2	1

2	1
4	3

2	1
4	1

3	2
4	1

3	3
2	1

3	4
4	3

2	2
3	4

Number Path

Copy out this puzzle. Working from one square to another, horizontally or vertically (never diagonally), draw paths to pair up each set of two matching numbers. No path may be shared, and none may enter a square containing a number or part of another path. Write your answers on a separate sheet of paper.

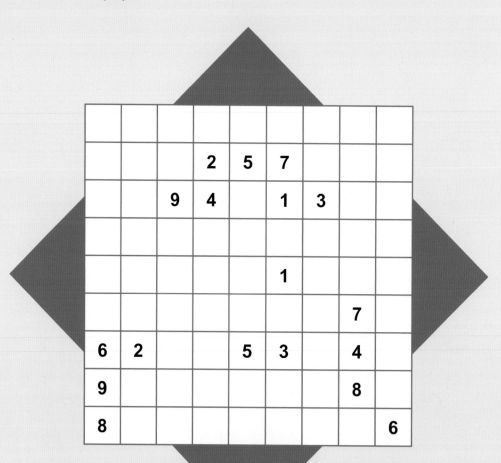

Hexagony

Can you place the hexagons into the grid, so that where any hexagon touches another along a straight line, the number in both triangles is the same? No rotation of any hexagon is allowed! Write your answers on a separate sheet of paper.

Mini Sudoku

Every row, column, and each of the four smaller boxes of four squares should contain a different number from 1 to 4 inclusive. Some numbers are already in place. Can you complete the grid? Write your answers on a separate sheet of paper.

13

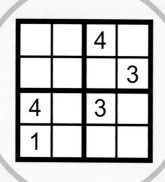

14

15

Every row, column, and each of the six smaller boxes of six squares should contain a different number from 1 to 6 inclusive. Some numbers are already in place. Can you complete the grid? Write your answers on a separate sheet of paper.

Making Arrangements

Arrange one each of the four numbers below, as well as one each of the symbols x (times), – (minus), and + (plus) in every row and column. You should arrive at the answer at the end of the row or column, making the calculations in the order in which they appear. Some are already in place. Write your answers on a separate sheet of paper.

4   6   7   9

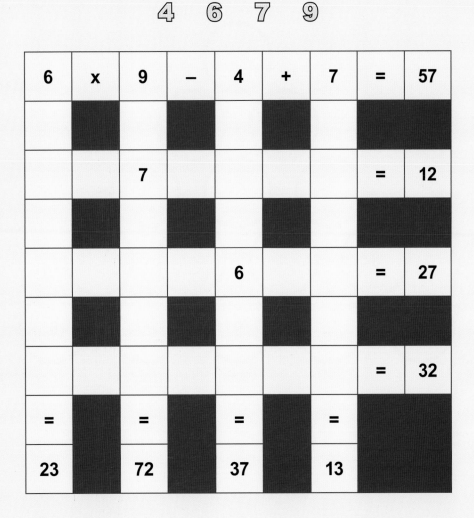

6	x	9	–	4	+	7	=	57
	■	7	■		■		■	=	12
	■		■	6			=	27
	■		■		■		=	32
=		=		=		=		
23		72		37		13		

Pyramid Plus

The number in each circle is the sum of the two numbers below it. Just work out the missing numbers in every circle! Write your answers on a separate sheet of paper.

17

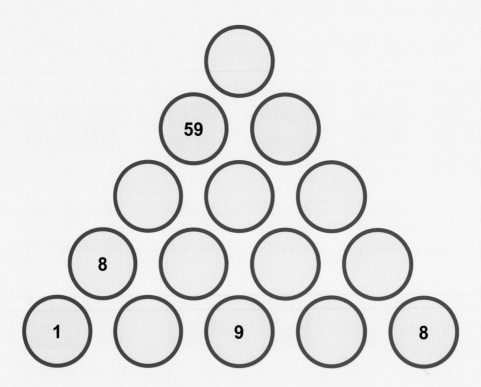

One to Nine

Using the numbers one to nine, complete these six equations (three reading across and three reading downward). Every number is used once only, and one is already in place. Write your answers on a separate sheet of paper.

18

1 2 3 4 5 6 7 8 9

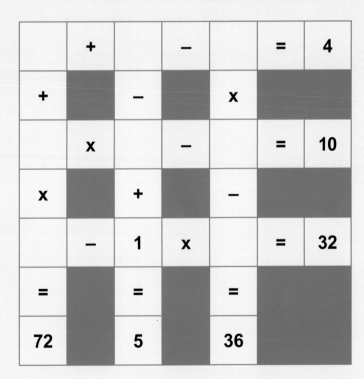

Number Crunch

Starting at the top left with the number provided, work down from one box to another, applying the mathematical instructions to your running total. Write your answers on a separate sheet of paper.

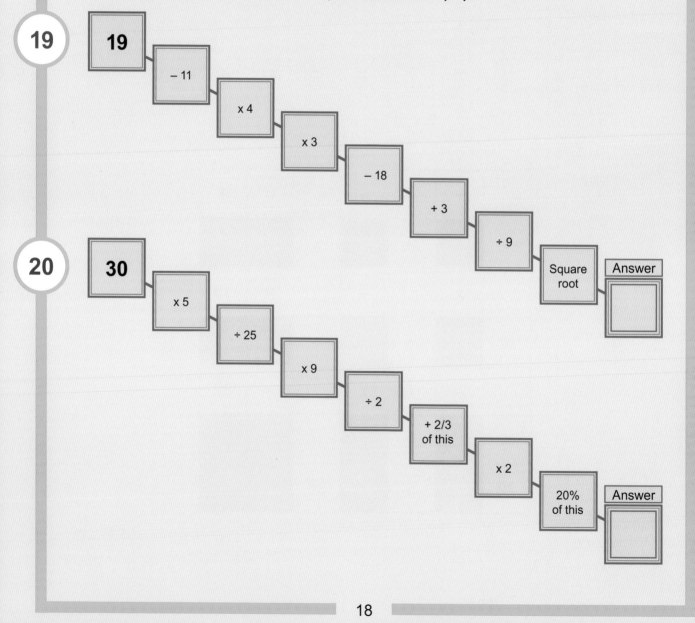

19

19 → − 11 → x 4 → x 3 → − 18 → + 3 → ÷ 9 → Square root → Answer

20

30 → x 5 → ÷ 25 → x 9 → ÷ 2 → + 2/3 of this → x 2 → 20% of this → Answer

Total Concentration

The blank squares below should contain whole numbers between 1 and 30 inclusive, any of which may occur more than once, or not at all. The numbers in every horizontal row add up to the totals on the right, as do the two long diagonal lines extending from corner to corner; those in every vertical column add up to the totals along the bottom. Write your answers on a separate sheet of paper.

							107
	27		19	12	26	19	133
	18	26	5		20	25	116
2		17	27	1	16		89
25	4	24		23		15	125
14	20	18	14	9	6		104
	28			17		8	107
11		11	29		30	15	133
95	131	120	111	88	144	118	100

Tile Twister

Place the eight tiles into the puzzle grid so that all adjacent numbers on each tile match up. Tiles may be rotated through 360 degrees, but none may be flipped over. Write your answers on a separate sheet of paper.

22

2	3				
1	3				

Tiles:

2	1
3	4

1	1
1	3

3	1
1	4

2	3
3	1

1	1
4	2

4	4
3	1

1	3
3	3

4	4
3	2

Pyramid Plus

The number in each circle is the sum of the two numbers below it. Just work out the missing numbers in every circle! Write your answers on a separate sheet of paper.

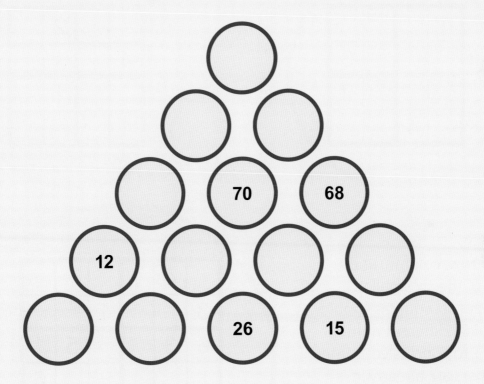

Mini Sudoku

Every row, column, and each of the four smaller boxes of four squares should contain a different number from 1 to 4 inclusive. Some numbers are already in place. Can you complete the grid? Write your answers on a separate sheet of paper.

24

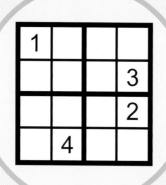

25

26

Every row, column, and each of the six smaller boxes of six squares should contain a different number from 1 to 6 inclusive. Some numbers are already in place. Can you complete the grid? Write your answers on a separate sheet of paper.

Total Concentration

The blank squares below should contain whole numbers between 1 and 30 inclusive, any of which may occur more than once, or not at all. The numbers in every horizontal row add up to the totals on the right, as do the two long diagonal lines extending from corner to corner; those in every vertical column add up to the totals along the bottom. Write your answers on a separate sheet of paper.

							47
11		2	15	29		5	78
30	21		4	24	1		101
	12	23	20		8	25	103
	28	10		1	13	7	84
7		19	14	17	26		102
	2	13		28	5	16	103
6	8	28	18			27	116
110	93	104	83	125	77	95	107

Hexagony

Can you place the hexagons into the grid, so that where any hexagon touches another along a straight line, the number in both triangles is the same? No rotation of any hexagon is allowed! Write your answers on a separate sheet of paper.

Number Path

Copy out this puzzle. Working from one square to another, horizontally or vertically (never diagonally), draw paths to pair up each set of two matching numbers. No path may be shared, and none may enter a square containing a number or part of another path. Write your answers on a separate sheet of paper.

29

4	5							7
			6					
			9					
		3			3			
4		2			2			
5		1			1			
7							9	6
8								8

Making Arrangements

Arrange one each of the four numbers below, as well as one each of the symbols x (times), – (minus), and + (plus) in every row and column. You should arrive at the answer at the end of the row or column, making the calculations in the order in which they appear. Some are already in place. Write your answers on a separate sheet of paper.

30

2   3   5   7

3	x	7	–	5	+	2	=	18
							=	24
x								
			x			7	=	14
				+				
							=	15
=		=		=		=		
9		28		34		30		

Hexagony

Can you place the hexagons into the grid, so that where any hexagon touches another along a straight line, the number in both triangles is the same? No rotation of any hexagon is allowed! Write your answers on a separate sheet of paper.

Pyramid Plus

The number in each circle is the sum of the two numbers below it. Just work out the missing numbers in every circle! Write your answers on a separate sheet of paper.

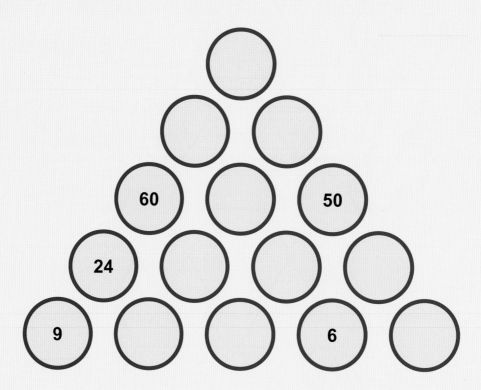

One to Nine

Using the numbers one to nine, complete these six equations (three reading across and three reading downward). Every number is used once only, and one is already in place. Write your answers on a separate sheet of paper.

33

1 2 3 4 5 6 7 8 9

	x		–	9	=	1
+		x		–		
	x		+		=	10
–		+		x		
	+		x		=	80
=		=		=		
3		9		48		

Mini Sudoku

Every row, column, and each of the four smaller boxes of four squares should contain a different number from 1 to 4 inclusive. Some numbers are already in place. Can you complete the grid? Write your answers on a separate sheet of paper.

34

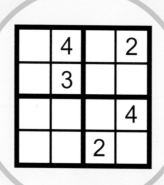

35

36

Every row, column, and each of the six smaller boxes of six squares should contain a different number from 1 to 6 inclusive. Some numbers are already in place. Can you complete the grid? Write your answers on a separate sheet of paper.

Number Crunch

Starting at the top left with the number provided, work down from one box to another, applying the mathematical instructions to your running total. Write your answers on a separate sheet of paper.

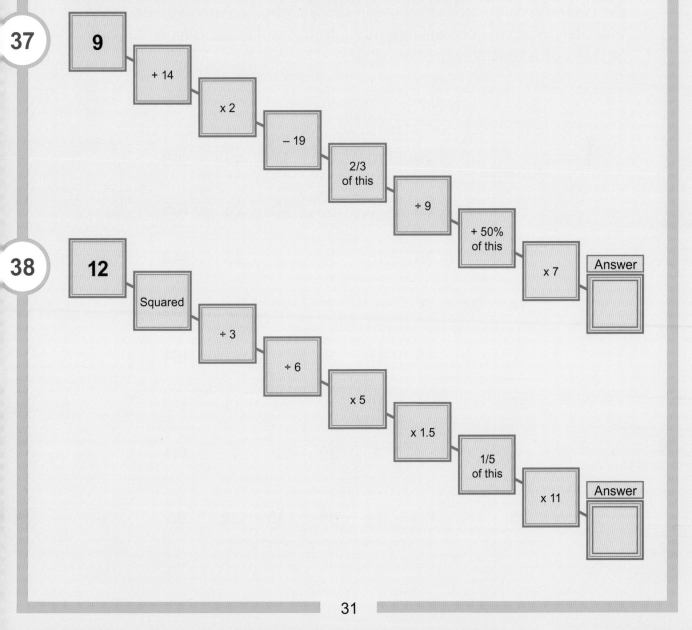

37

9

+ 14

x 2

− 19

2/3 of this

÷ 9

+ 50% of this

x 7

Answer

38

12

Squared

÷ 3

÷ 6

x 5

x 1.5

1/5 of this

x 11

Answer

Total Concentration

The blank squares below should contain whole numbers between 1 and 30 inclusive, any of which may occur more than once, or not at all. The numbers in every horizontal row add up to the totals on the right, as do the two long diagonal lines; those in every vertical column add up to the totals along the bottom extending from corner to corner. Write your answers on a separate sheet of paper.

39

								94
	27	25	10		17	13	**118**	
	3	24	16		10	23	**109**	
18		30	16	19		23	**144**	
14	24	5		2	26		**111**	
4	15	15	27	9			**101**	
28	12		20		21	25	**113**	
			8	20	22	7	**111**	
98	**112**	**121**	**108**	**95**	**135**	**138**	**90**	

Tile Twister

Place the eight tiles into the puzzle grid so that all adjacent numbers on each tile match up. Tiles may be rotated through 360 degrees, but none may be flipped over. Write your answers on a separate sheet of paper.

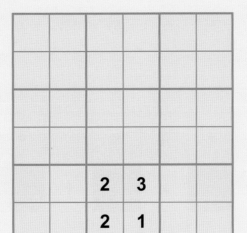

2	2
3	3

2	2
4	1

2	4
1	2

2	3
4	2

3	3
1	4

2	2
2	1

3	4
4	2

4	3
2	1

One to Nine

Using the numbers one to nine, complete these six equations (three reading across and three reading downward). Every number is used once only, and one is already in place. Write your answers on a separate sheet of paper.

1 2 3 4 5 6 7 8 9

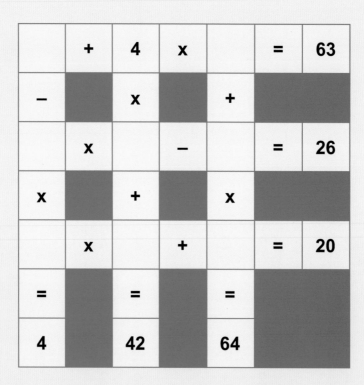

Making Arrangements

Arrange one each of the four numbers below, as well as one each of the symbols x (times), – (minus), and + (plus) in every row and column. You should arrive at the answer at the end of the row or column, making the calculations in the order in which they appear. Some are already in place. Write your answers on a separate sheet of paper.

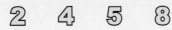

2 4 5 8

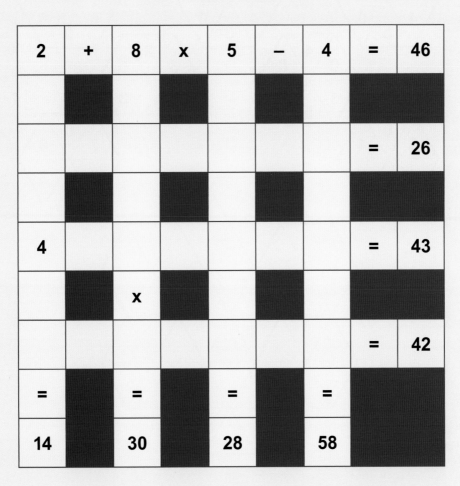

Hexagony

Can you place the hexagons into the grid, so that where any hexagon touches another along a straight line, the number in both triangles is the same? No rotation of any hexagon is allowed! Write your answers on a separate sheet of paper.

43

Number Path

Copy out this puzzle. Working from one square to another, horizontally or vertically (never diagonally), draw paths to pair up each set of two matching numbers. No path may be shared, and none may enter a square containing a number or part of another path. Write your answers on a separate sheet of paper.

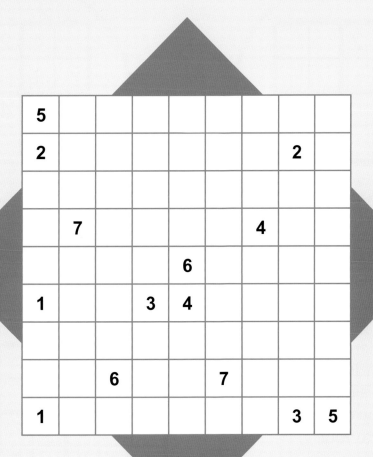

Mini Sudoku

Every row, column, and each of the four smaller boxes of four squares should contain a different number from 1 to 4 inclusive. Some numbers are already in place. Can you complete the grid? Write your answers on a separate sheet of paper.

45

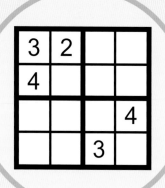

46

47

Every row, column, and each of the six smaller boxes of six squares should contain a different number from 1 to 6 inclusive. Some numbers are already in place. Can you complete the grid? Write your answers on a separate sheet of paper.

Word Puzzles

48 ### The Expensive Refund

An elegant woman walks into a jewelry store and looks through the selection of earrings. She tries some on and eventually chooses a pair costing $250. She pays in cash and leaves.

Several minutes later she sees her reflection in a store window; she stops and looks more closely at the new earrings and decides they're not quite right after all. The woman returns to the store and remembers a pair that she had seen earlier. She finds the salesman and explains that she would like to return the first pair and now have the second.

"That's fine ma'am," he says, as she takes the box with the second pair and hands him the first. "This second pair costs $500, so I'll have to charge you $250."

"Oh, no," the woman replies. I've given you $250 already and now I'm handing you earrings worth another $250. That makes $500, so we're even."

And with that she walked out, leaving the salesman puzzled. Can you figure it out?

49 ### Two Much Trouble

As a punishment for talking in class, Ella had to write down all the numbers from 1 to 100. How many times did she write the number "2"?

50 ### The Cost of Living

Inflation is a word used to describe how much prices rise over a certain period, usually a year (annual inflation rate). If the annual rate of inflation is 5 percent, and a half gallon of ice cream costs $3.50 now, how much will it cost after three years?

Solutions

1

95 + 18 = 113, 113 − 72 = 41, 41 x 2 = 82, 82 x 1.5
= 123, 123 − 16 = 107, 107 − 17 = 90, 90 x 4 = 360

2

2^2 = 4, 4 x 9 = 36, square root of 36 = 6, 6 x 7 =
42, 42 ÷ 3 = 14, 14 + 8 = 22, 22 + 38 = 60

3

8	−	1	x	5	=	35
+		+		x		
2	x	9	+	3	=	21
−		x		+		
4	x	6	−	7	=	17
=		=		=		
6		60		22		

4

5	+	9	x	3	−	6	=	36
−		x		x		+		
3	x	5	−	6	+	9	=	18
x		−		+		x		
9	+	6	x	5	−	3	=	72
+		+		−		−		
6	−	3	x	9	+	5	=	32
=		=		=		=		
24		42		14		40		

5

138

18	18	3	24	2	6	22	93
2	3	17	11	6	15	11	65
12	23	19	10	13	5	29	111
12	9	10	30	19	25	14	119
7	26	17	13	1	8	4	76
4	14	1	5	16	15	8	63
27	20	16	7	9	21	28	128

82	113	83	100	66	95	116	114

6

3	2	1	4
4	1	2	3
1	3	4	2
2	4	3	1

7

3	2	1	4
1	4	3	2
2	1	4	3
4	3	2	1

8

5	4	2	6	3	1
6	3	1	2	5	4
1	2	5	4	6	3
3	6	4	1	2	5
4	5	6	3	1	2
2	1	3	5	4	6

9

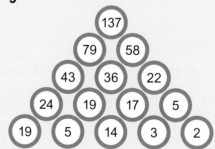

10

3	3	3	1	1	1
2	1	1	2	2	4
2	1	1	2	2	4
4	3	3	2	2	3
4	3	3	2	2	3
3	4	4	1	1	1

11

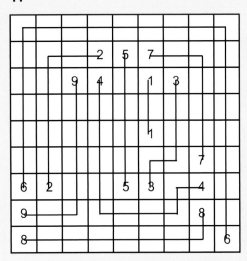

12

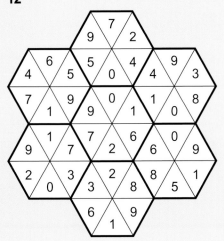

13

3	1	4	2
2	4	1	3
4	2	3	1
1	3	2	4

14

3	2	4	1
4	1	3	2
2	3	1	4
1	4	2	3

15

5	4	6	3	1	2
3	2	1	5	4	6
6	5	4	2	3	1
2	1	3	6	5	4
4	6	5	1	2	3
1	3	2	4	6	5

16

6	x	9	−	4	+	7	=	57
−		+		x		−		
4	+	7	−	9	x	6	=	12
x		−		−		x		
7	−	4	x	6	+	9	=	27
+		x		+		+		
9	+	6	−	7	x	4	=	32
=		=		=		=		
23		72		37		13		

17

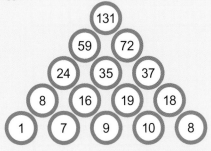

18

2	+	7	−	5	=	4
+		−		x		
6	x	3	−	8	=	10
x		+		−		
9	−	1	x	4	=	32
=		=		=		
72		5		36		

19

19 − 11 = 8, 8 x 4 = 32, 32 x 3 = 96, 96 − 18 = 78, 78 + 3 = 81, 81 ÷ 9 = 9, square root of 9 = 3

20

30 x 5 = 150, 150 ÷ 25 = 6, 6 x 9 = 54, 54 ÷ 2 = 27, 27 + 18 = 45, 45 x 2 = 90, 20% of 90 = 18

21

2	x	5	−	9	=	1
+		x		−		
7	x	1	+	3	=	10
−		+		x		
6	+	4	x	8	=	80
=		=		=		
3		9		48		

22

2	3	3	3	3	2
1	3	3	1	1	3
1	3	3	1	1	3
1	1	1	4	4	4
1	1	1	4	4	4
4	2	2	3	3	2

23

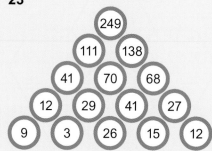

24

1	3	2	4
4	2	1	3
3	1	4	2
2	4	3	1

25

3	1	2	4
4	2	3	1
2	4	1	3
1	3	4	2

26

2	4	1	6	3	5
3	5	6	1	4	2
6	3	2	4	5	1
4	1	5	3	2	6
5	6	4	2	1	3
1	2	3	5	6	4

27

47

11	6	2	15	29	10	5	78
30	21	9	4	24	1	12	101
4	12	23	20	11	8	25	103
22	28	10	3	1	13	7	84
7	16	19	14	17	26	3	102
30	2	13	9	28	5	16	103
6	8	28	18	15	14	27	116

110	93	104	83	125	77	95	107

28

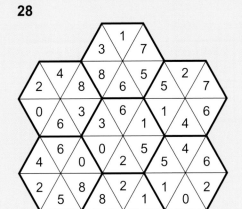

29

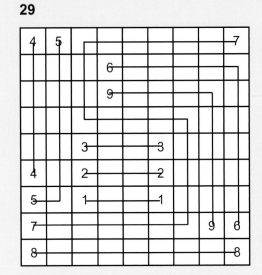

30

3	x	7	−	5	+	2	=	18
+		−		x		+		
5	−	2	x	7	+	3	=	24
x		x		−		x		
2	+	5	x	3	−	7	=	14
−		+		+		−		
7	+	3	x	2	−	5	=	15
=		=		=		=		
9		28		34		30		

31

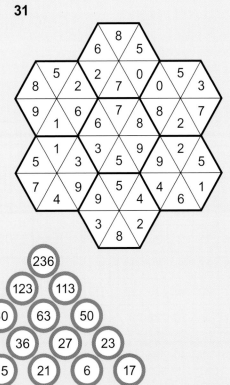

32

236 / 123 113 / 60 63 50 / 24 36 27 23 / 9 15 21 6 17

43

Solutions

33

2	x	5	−	9	=	1
+		x		−		
7	x	1	+	3	=	10
−		+		x		
6	+	4	x	8	=	80
=		=		=		
3		9		48		

34

1	4	3	2
2	3	4	1
3	2	1	4
4	1	2	3

35

1	2	4	3
3	4	2	1
2	1	3	4
4	3	1	2

36

5	4	1	2	6	3
6	2	3	5	4	1
2	3	5	4	1	6
4	1	6	3	2	5
1	5	2	6	3	4
3	6	4	1	5	2

37

9 + 14 = 23, 23 x 2 = 46, 46 − 19 = 27, 27 ÷ 3 x 2 = 18, 18 ÷ 9 = 2, 2 + 1 = 3, 3 x 7 = 21

38

12^2 = 144, 144 ÷ 3 = 48, 48 ÷ 6 = 8, 8 x 5 = 40, 40 x 1.5 = 60, 60 ÷ 5 = 12, 12 x 11 = 132

39

94

9	27	25	10	17	17	13	118
11	3	24	16	22	10	23	109
18	12	30	16	19	26	23	144
14	24	5	11	2	26	29	111
4	15	15	27	9	13	18	101
28	12	1	20	6	21	25	113
14	19	21	8	20	22	7	111
98	112	121	108	95	135	138	90

40

1	2	2	1	1	2
3	4	4	2	2	2
3	4	4	2	2	2
4	2	2	3	3	3
4	2	2	3	3	3
1	2	2	1	1	4

41

5	+	4	x	7	=	63
−		x		+		
3	x	9	−	1	=	26
x		+		x		
2	x	6	+	8	=	20
=		=		=		
4		42		64		

42

2	+	8	x	5	−	4	=	46
x		−		−		+		
5	x	4	−	2	+	8	=	26
−		+		x		x		
4	+	2	x	8	−	5	=	43
+		x		+		−		
8	x	5	+	4	−	2	=	42
=		=		=		=		
14		30		28		58		

43

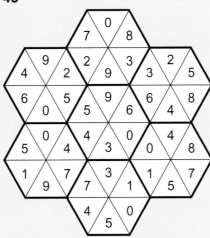

44

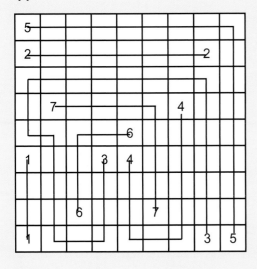

45

3	2	4	1
4	1	2	3
2	3	1	4
1	4	3	2

46

3	4	2	1
2	1	3	4
1	2	4	3
4	3	1	2

47

4	2	1	5	3	6
6	5	3	4	2	1
5	6	2	1	4	3
3	1	4	2	6	5
2	3	5	6	1	4
1	4	6	3	5	2

48

A purchase is an exchange, or a balance like a seesaw. The buyer exchanges money for a product. If the product goes back to the seller, then the money must go back to the buyer. The elegant woman is trying to use the $250 twice.

49

19 times

50

Just over $4.05

Glossary

adjacent	Close to or—more commonly—next to.
calculation	The use of math to find a solution.
column	A line of objects that goes straight up and down.
concentration	Thinking very hard and examining every possibility.
diagonal	Moving in a slanted direction, halfway between straight across and straight down.
grid	A display of crisscrossed lines.
hexagon	A six-sided object.
horizontal	A direction that is straight across.
inclusive	Including both ends of a series (two to five inclusive means 2, 3, 4, and 5).
matching	Exactly the same as.
mini	Small (an informal word).
occur	To happen.
pyramid	A triangular shape with one side level to the ground and a point at the top.
rotate	Travel in a circular motion.
rotation	Circular motion.
row	A line of objects that goes straight across.
shared	Having something the same as something else.
square root	A number that, if multiplied by itself, produces the original number (3 is the square root of 9; 4 is the square root of 16).
squared	When a number is multiplied by itself (3 squared = 3 x 3 = 9).
vertical	A direction that is straight up and down.
whole number	A number that has no decimals (4 is a whole number; 4.3 is not a whole number).

Further Information

For More Information

Consortium for Mathematics (COMAP)
175 Middlesex Turnpike, Bedford, MA 01730
(800) 772-6627 http://www.comap.com/index.html
COMAP is a nonprofit organization whose mission is to improve mathematics education for students of all ages. It works with teachers, students, and business people to create learning environments where mathematics is used to investigate and model real issues in our world.

MATHCOUNTS Foundation
1420 King Street, Alexandria, VA 22314
(703) 299-9006 https://mathcounts.org/sslpage.aspx
MATHCOUNTS is a national enrichment, club, and competition program that promotes middle school mathematics achievement. To secure America's global competitiveness, MATHCOUNTS inspires excellence, confidence, and curiosity in U.S. middle school students through fun and challenging math programs.

National Council of Teachers of Mathematics (NCTM)
906 Association Drive, Reston, VA 20191-1502
(703) 620-9840 http://www.nctm.org
The NCTM is a public voice of mathematics education supporting teachers to ensure equitable mathematics learning of the highest quality for all students through vision, leadership, professional development and research.

Web Sites

Due to the changing nature of Internet links, Rosen Publishing has developed an online list of Web sites related to the subject of this book. This site is updated regularly. Please use this link to access this list:

> http://www.rosenlinks.com/msw/prob

Further Reading

Abramson., Marcie F. *Painless Math Word Problems.* New York, NY: Barron's Educational Series, 2010.

Ball, Johnny. *Why Pi?* New York, NY: Dorling Kindersley, 2009.

Fisher, Richard W. *Mastering Essential Math Skills: Pre-Algebra Concepts: 20 Minutes a Day to Success.* Los Gatos, CA: Math Essentials, 2008.

Shortz, Will et al. *I Can KenKen! 75 Puzzles for Having Fun with Math.* New York, NY: St. Martin's Griffin, 2008.

Index